# The Magical Life Cycle of a Butterfly

**Miss Sierra**

In the garden, flowers bloom,
Butterflies now find their room.

Once, they started very small,
Here's the journey, here's it all!

Eggs are laid upon the leaves,
Tiny, round, beneath the trees.

From the egg, the hatchling came,
Caterpillar is its name.

Munching, crunching, all day long,
Eating leaves till it grows strong.

Growing big with every snack,
Soon it's ready, no more lack.

In a chrysalis, it sleeps,
Safe and sound,
and snug it keeps.

Through the days, a change begins,
Wings will grow and strength within.

From the shell, so soft and bright, Wings unfold into the light.

Butterfly now takes its flight,
Dancing on the breeze, what sight!

Flowers bloom, and nectar sweet,
Butterfly lands on its feet.

Once it crawled and ate all day,
Now it soars and flies away.

Now it's time to start again,
Eggs are laid just like back then.

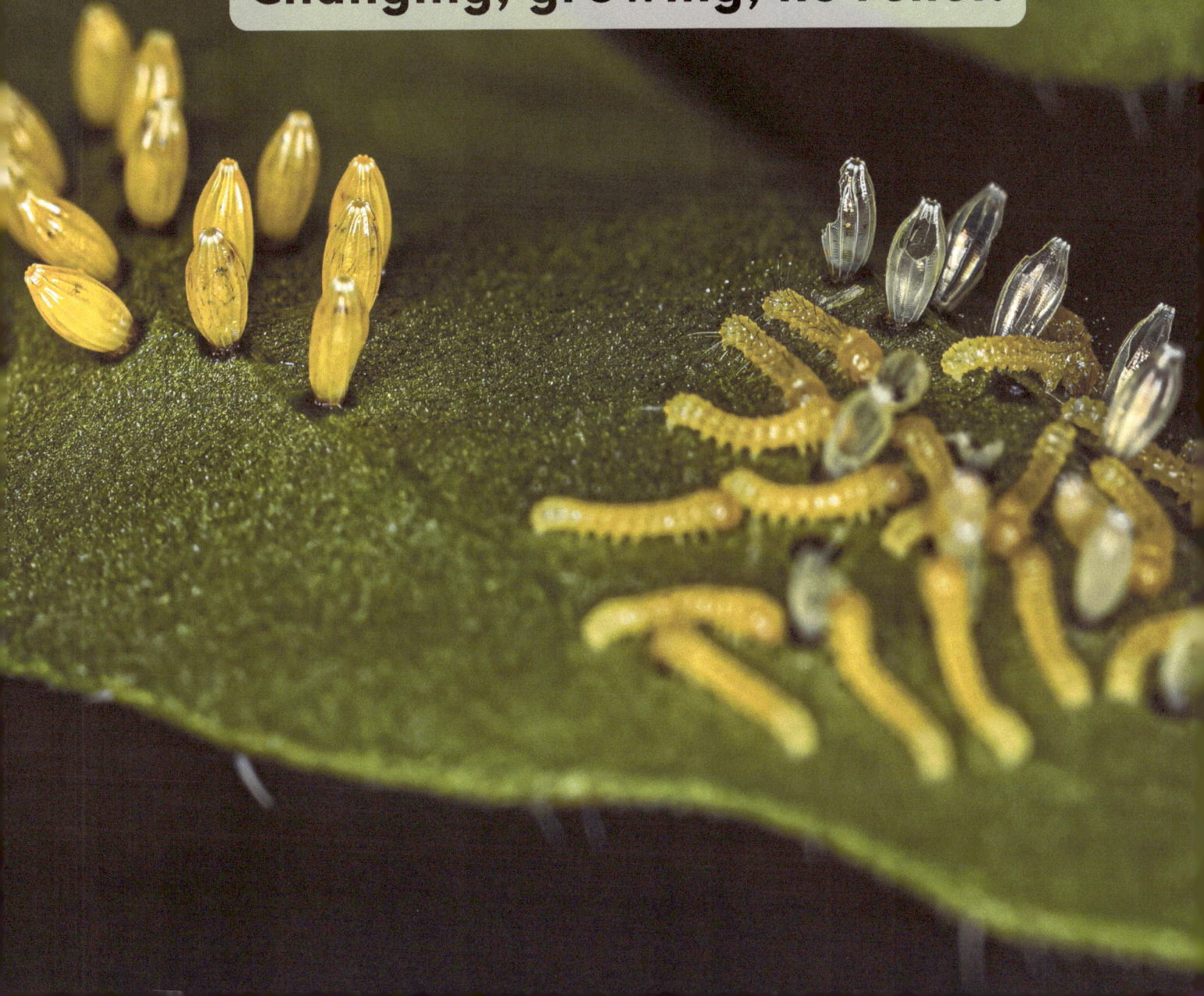

Life begins upon the leaf,
Changing, growing, no relief!

Watch them flutter, bright and free,
Nature's magic, you will see.

In a cycle, round they go,
Life keeps growing, watch it flow!

# FACTS

- A butterfly's life cycle has four stages: egg, caterpillar (larva), chrysalis (pupa), and adult butterfly.

- Butterfly eggs are very small and are usually laid on the underside of leaves to keep them safe.

- After eating and growing enough, the caterpillar forms a hard shell called a chrysalis, where it will change into a butterfly.

- As caterpillars grow, they shed their skin several times. This process is called molting.

- Butterflies play an important role in pollination. As they move from flower to flower to drink nectar, they transfer pollen, helping plants reproduce.

- Inside the chrysalis, the caterpillar body completely restructures, forming the wings, antennae, and legs of the adult butterfly.

- The amount of time in the chrysalis varies depending on the species and environmental conditions, ranging from a few days to several weeks.

- When a butterfly comes out of the chrysalis, its wings are soft and crumpled. It has to wait for its wings to dry before it can fly.

- Many caterpillars and butterflies use camouflage or mimicry to protect themselves from predators.

- Most butterflies only live for a few weeks after they become adults, though some species, like the monarch, can live for several months.

# VOCABULUARY

🔶 **Metamorphosis** - The process of transformation from one stage of life to another, like when a caterpillar becomes a butterfly.

🔶 **Egg** - The first stage of a butterfly's life cycle, where a tiny caterpillar grows inside.

🔶 **Caterpillar (Larva)** - The second stage of the butterfly's life cycle. The caterpillar eats leaves and grows bigger before changing into a chrysalis.

🔶 **Chrysalis (Pupa)** - The third stage of the life cycle, where the caterpillar transforms into a butterfly inside a hard shell.

🔶 **Nectar** - A sweet liquid made by flowers that butterflies drink for food.

# YOUR COMMUNITY PARTNER
## SiohanPress.com

SCAN ME

SIOHAN PRESS

A PUBLISHING COMPANY

Mailing List

Together we increase the confidence and abilities of the children we love!

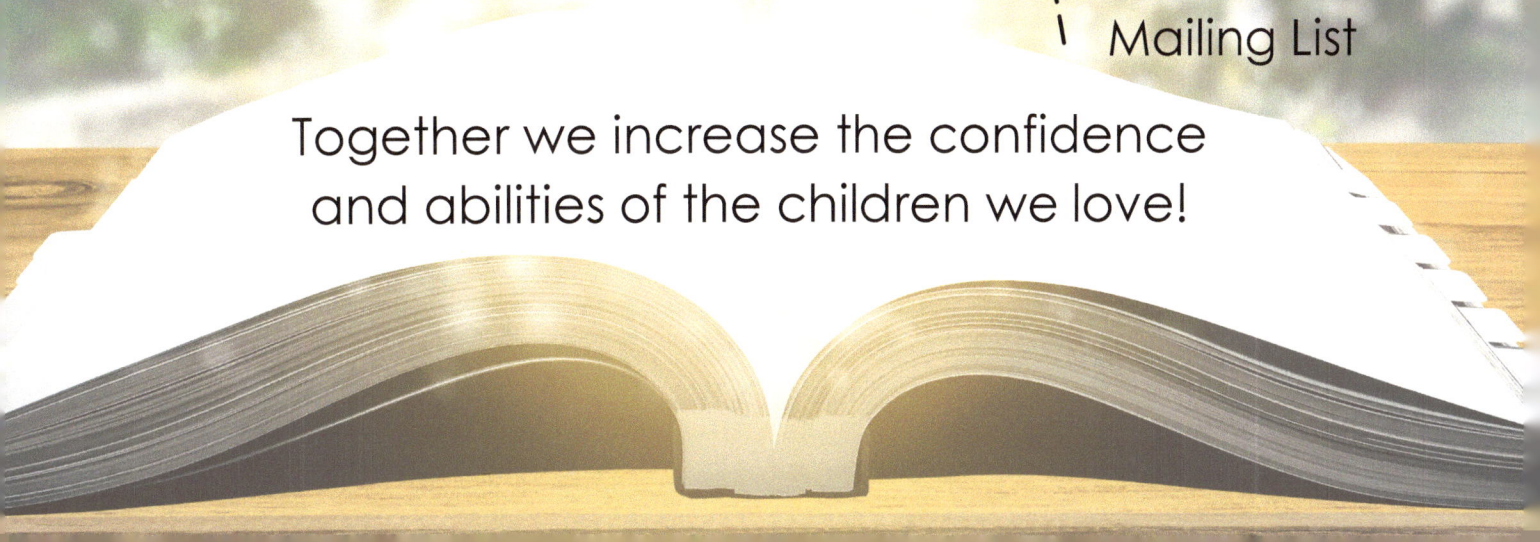

www.ingramcontent.com/pod-product-compliance
Lightning Source LLC
Chambersburg PA
CBHW061147030426
42335CB00002B/143